Yes, we can unify Quantum Mechanics
and Relativity

ISBN: 1-4563-1244-8
ISBN-13: 9781456312442

Yes, we can unify Quantum Mechanics and Relativity

Theory of Reigning Element

Prasad Vemulapalli

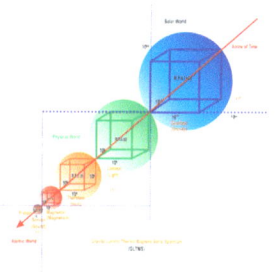

2010

Contents

Preface

It all began on that fateful evening in 1992 when I was sitting in my backyard watching a beautiful sunset, I saw some luminous spheres dancing before my eyes it seemed they were in perennial Brownian motion, I wondered what these could be. I fondly called them 'Helumes', combining Helios (meaning Sun god) and Luminous (meaning light) as I thought these were packets of energy carrying light from sun and then suddenly a strange thought occurred to me why can't we think of the sun as Nucleus containing Proton/s and earth as electron thus my quest for viable design for our solar system began and during past 18 years many mysterious concepts unfolded in my mind and a whole new gamut of science evolved. Now I wonder how I could so aptly name them as indeed they are the ions of beryllium, the least quanta of energy from sun which contain 10^2 joules of energy which I renamed later as MagnetIons for reasons that would be explained in my theory.

Whenever my theory is implemented the science of physics as we know would change forever and better and the life of physicists all over will become far easier and simpler.

We will need not wreck our brains with the complicated equations anymore everything will be simple and straight forward as it used to be in Newton's time and I fear one more Einstein may prove me wrong but until then I hope our physical world would be far simpler.

It is simply a world of squaring masses and cubing energies related by $M^2 = E^3$

First of all on this great momentous scientific occasion I wish to remember, profusely thank and honor the memory my scientific fraternity starting with Newton, Kepler, Bohr, Pauli, Pascal, Mendeleev, Fleming, Maxwell, Planck and Einstein whose theories I used to develop my own, other great scientists like Pythagoras, Galileo, Copernicus, Aristotle who have suffered(in pursuit of knowledge some have even laid their lives in the hands of wicked roman emperors) to make our lives better and of course it would be monstrous of me not to mention Dr. .Stephen Hawking, whose book ' a brief history of time' inspired me to come up with an alternative to unify the quantum theory and the theory of relativity, I feel really very glad and lucky to have solved the problem of unification while Dr. Hawking is still around because he is the only legend living whose beautiful mind alone can understand the implications of my theory. I miss Dr. Einstein whom I revere as god of science, I would have liked the flamboyant Carl Sagan and Arthur C Clarke to be around on this occasion as they would have enjoyed this like anything.

I would like to specially remember Nicola Tesla, the forgotten genius (courtesy Edison), may his soul rest in peace at last.

I would also like to remember Igor Kurchatov who almost single handedly built Hydrogen bomb for Russia.

Among the late artists I would like to remember Telugu Actor Annagaru Nandamuri Taraka Rama Rao, Babaigaru Gummadi Vekateshwara Rao, Babaigaru S.V. Ranga Rao, Michael Jackson, Andy and Maurice Gibbs of Bee-Gees whose art I admired , I also would like to thank living legend Annagaru Akkineni Nageshwara Rao for providing clean entertainment for over 60 years and still going strong

During this nostalgic journey through space, every milestone was equally important and was a great achievement but if you ask my personal choice I would definitely choose Einstein's inventing time as the single most critical scientific invention of all time as thinking of abstract thing requires greatest creativity.

I consider Newton, Planck and Einstein, my Trinity of science gods.

I should frankly admit that my invention was nothing but sheer serendipity and I cannot be compared with the any of the great intellectuals of the yore, I am Chief Architect Jacques Clouseau of Science I did whatever I did through by pure intuition.

Famous Indian news paper columnist Khushwanth Singh once wrote about Jats, a warrior community from North West India, that their heart dictates first and mind justifies later, I think I am no different (In fact I am also a Chowdary from South), I follow my heart it may mislead sometimes but if it is pure enough you will certainly reach your destiny, My life being best example.

I must admit my math is very poor, it is so poor that in fact I failed to clear math paper in my 7th standard mathematics exam, only with grace marks I could barely clear the exam.

I am not at all good at execution details but surely am very good at conceptual design.

I think more of myself as St. Moses to whom the real god revealed himself and wrote Ten Commandments on rock with fire when the people around him started squabbling over the fabric of universe as to whether it is made up of discrete quanta or strings, I found both were right, but only few specific conditions apply

I would also like to be remembered as the guy who unified the theories of Newton, Planck and Einstein.

I have a very clumsy unorthodox way of stating things, I request you to bear with me and consider only the verity of the statements I make , fruits of your patience will definitely be rewarded I assure you.

I humbly submit my work for your review and validation.

'To err is human, to forgive is divine'—William Shakespeare.

I believe in intuition and inspiration. Imagination is more important than knowledge. For knowledge is limited, where as imagination embraces the entire world, stimulating progress, giving birth to evolution. It is, strictly speaking, a real factor in scientific research, a new type of thinking is essential if mankind is to survive and move toward higher levels.—Albert Einstein

I am sure I could have gone wrong somewhere but the greatest of the scientists, Newton also went wrong so I am just doing my duty fit for this particular event of my lifetime in space.

So far it was thought that we cannot combine quantum theory with relativity but Yes, we can, and I did it, read through the paper how small a tweak I had to give to the theory of relativity to change it all.

They used to refer to the great October revolution as ten days that shook the world, about my theory I would say sixty four pages that changed the whole known physical reality.

Rest assured none of the existing laws have been violated or disturbed; only they have been interpreted in a different perspective.

At first glance, my theory may even seem to border on Ludicrousness and abrupt but please bear with me, I promise you will enjoy the simplicity and intensity slowly.

It's only simple words and those are all I have to show you the whole reality.

No complex equations have been used, even a lay man can understand what is being explained, and in fact that was the basic purpose of this book, teaching the most complex theories in layman's terms.

I did my share of cobbler's work as mainframe programmer in various organizations from 1998 to 2010 but even it bid me good bye when I could not keep in blue books of manager one Mr. Ram Iyengar, Whom I should thank but for him who got me removed from my cobbler post, this theory would never have seen the light of the day.

In essence Newton invented gravity and physical laws for R.F.3, Planck invented Photon, Einstein invented the R.F.4 or Time and I just had to unify them all and discover the Gravitlon and invent Viton.

After I almost completed my theory, I happened to read about Burkhard Heim's theory of everything (as usual I did not understand the math part of it), first of all I bow my head in reverence to that great genius who I gather had lost both hands in war and even ears but still could retain indomitable scientific spirit to come up with such grand theory, may his soul rest in peace. He rightly deserves to be called the Stephen Hawking of 20th century.

Now coming to his theory I am very impressed with his mass formulae, however there are few things that I do not agree with in Heim's theory, for example the very concept of six dimensional metron, in my view though there are six energy dimensions, there are only five mass dimensions and sixth mass dimension is just the holistic representation of the combination of physical three dimensions, hypothetical fourth dimension Time and fifth dimension the gravitophoton ($3^3+4^3+5^3 = 6^3$). The metron Heim talks about cannot be just a small microscopic indivisible particle of six dimensions but should vary in size over the different worlds, however there can be one metron of finite size for each of the worlds, for example C^6 for solar world. The math is fine as he employed R6 metric. I was really astonished at his prediction that Gravitlon can be artificially created and yes, we can theoretically produce Gravitlon by clubbing two Photlons.

After I have proof read my book and when it was almost ready to be printed, I read about the Kaluza Klein theory using five dimensions, fifth of which exactly describes electromagnetic phenomenon, according to my theory what Gravity is to our physical world Magnetism is to Atomic world and what Light is to our world Electricity is to the atomic world and this is the reason why the Newton's law of Gravity and Coulomb's law of Electricity are so similar.

All the while I was really bothered and even felt guilty about not being able to provide the supporting math but after I read their theory I felt greatly relieved as after all the math is already in place and I need not wreck my brain to come up with necessary math. I fully endorse the Kaluza Klein theory and subsequent theories based on it like String theory with 11 dimensions, however I do not subscribe to the absurd concept of parallel worlds in String theory. Just in case people think that I am opposed to classical standard Quantum Mechanics I want to clarify that I completely believe in QED and greatly admire the work of likes of Neils Bohr, Richard Feynman.

Acknowledgements

Of course I would like to thank my family(my father Late Vemulapalli Venkateswara Rao, mother Vemulapalli Seetha Ravamma, Brother Late Vemulapalli Kishore Babu, Sister in law Vemulapalli Vasantha, Nephew Vemulapalli Varun, Sister Kilaru Kusuma Devi, Brother in law Late Kilaru Vijay Kumar, Nephew Kilaru Sarath Chandra, Kilaru Ravi Theja, Sister Kurra Krishna Kumari,Brother in law Kurra Ravindra Babu, Niece Kurra Mounika, Nephew Kurra Pratheek, Sister Late Vemulapalli Rajani Kumari, Sister Karipineni Bhavani Devi, Brother in law Karipineni Sireesh Kumar, Nephew Karipineni Harshavardhan, Nephew Karipineni Sidharth, and my wife Vemulapalli Padmaja), friends(Masa Damodar, Meduru Sridhar, Rajneesh, Ravi, Pankaj, Chakri, John, Disney, Bison, Vasulu, Prasad babu, Nani, Guddanti, Talasila Suresh Chowdary, Suman, Chari, Butla Srihari, Bilakanti Sreenivasulu, Guduru Rama Krishna, Sheik Shabbar Ahmed, Mallikarjun Palepu, Chandrasekhar Bagoor, Bommineni Mohan, Ganga Siva Rama Krishna, Vemparala Ujjwal, Yalamanchili Chakravarthy, Chalasani Ajay, Chalasani Kalyan,Maddukuri Ramakrishna, Kavati Arun, Fons Cluitmans, Hans Okkerman, Henk Van De Bor, Hein Jurg, Sajeeth Nair, Ravindra Gharpure, Bhushan Renuse, Kanakamedala Sudhakar Rao, Khaleel Ahmed Khan, Sridhar Raju, Srini Basabathina, Rajasekhar Gorla, George Putnam, Arthur Scoby, Javed, Pochampalli Srinivas, Eadara Srinivas, Harsh Walia, Devender Mann, Abhijeet Kulkarni, Satyajit Mishra, Inturi Srinivas, Ede Gopal and Priyadatha), relations(Kongara Parthasarathi, Sriramoju Laxmi Narayana, Manohar Rao, Yarlagadda Siva Prasad, Yarlagadda Sudheer Babu, Karipineni Rajasekhar, Chalasani Venkata Ram Kumar) and our family's well wisher Kopparthi Nageswara Rao, who stood by me through thick and thin.

I would like to dedicate this work to mentor, philosopher, and guide of my childhood days, Annayya Gaddipati Babu Rajendra Prasad, who presented me with biography of Dr. Einstein and thus roused interest in me to know more about science.

I want to thank user RealityCheck of www.Physforum.com who proved to be the only sensible guy who encouraged people who could think out of the cube and did some useful thinking himself while rest were frogs in their confounded cube and merely indulged in useless trolling and baiting, I sincerely pray for his health and hope that he would contribute something really meaningful to the science of physics that we all physicists love.

Last but not least, I would also like to thank users Dhcracker of www.Physforum.com and Drowsy Turtle and Ophiolite of www.thescienceforum.com, who have provided constructive criticism but for which the major flaws of my theory could not have been corrected.

This edition being the first there are bound to be many errors, both unintended Logical as well as Typographical errors, please point out any such errors so that they can be rectified in the next edition

The ultimate aim of this book is to contribute something useful to the science of physics, so constructive criticism or suggestions with the aim of improving my theory are always welcome and would be appreciated.

If any of you have any valid questions or useful feedback you can always reach me at site www.prasadvrk.com.

I would specially like to thank my wife who designed the illustrations to my satisfaction.

I would like to thank Heera Bai of Harika Graphics for providing the high resolution images.

Finally I would like to thank Wikipedia which made knowledge which was so far hidden away behind secluded walls of few chosen scholars available to the public.

This book is a tribute to the freedom loving people of the biggest democracy, the U.S.A. who had the temerity to elect an African American as president which would not have been possible anywhere else
So enjoy your journey to the next energy dimension through time.

Chapter 1
Basic Math

Before we proceed with the Theory of Reigning Element let us first refresh our basic geometry and calculus a bit, which is essential to understand what follows

Geometry of Circle:

Circumference of Circle:
If we integrate pi w.r.t. radius, we get Integral pi dr = pi *r
The circumference of a circle is 2 * pi*r, where r is the radius of circle

Area of Circle:
If we integrate circumference w.r.t. radius then Integral 2*pi*r dr = 2*pi*r^2/2 or pi*r^2, which is nothing but the area of circle

Volume of Sphere:
Let us see what we get if we integrate the surface area of a sphere which is 4 *pi*r^2
integral 4*pi*r^2 dr =4/3 *pi*r^3, which is nothing but the volume of a sphere

Volume of Cone:
If we consider cones of base 2r and angle 90 Degrees then Volume of cone = 1/3*pi*r^2*r = 1/3*pi*r^3

Volume of Cylinder:
Volume of Cylinder = 1//3 *pi*r^2x h, if we consider a cylinder of height r then Volume = 1/3 *pi*r^3,

Volume of line of spheres:
Let us proceed even further and try integrating the volume of a sphere
6 * Integral 4/3 *pi*r^3 dr = 24/3*pi* r^4/4 = 2*pi* r^4 = 2*pi* (r^2)^2, i.e. if we integrate a sphere we are getting back to the area of a circle whose radius is r^2, which means that the fourth dimension is nothing but square of second dimension, which is exactly what I proposed in Part1 and marks the beginning of Euclidean geometry
In other words we can say a circle is stack of radii of length r placed side by side radially all around 360 degrees and sphere is stack of such circles placed radially all around 360 x 360 degrees

We know pi radians = 180 degrees

Geometry of Cube:

Line: Line is linear stack of points of unit length

Length of a line:
Length of a line = a , where a is the number of unit points

Square: Square is stack of lines so the area of a square is a^2
If we integrate a line of length a , integral a da = a^2 / 2, which means we need to integrate two lengths to get one perfect square

Circumference of Square:
Circumference of a square of side a = 4a

Cube: Cube is stack of squares of area a^2 over a height of length a
If we integrate square of area a^2, integral a^2 da = a^3 / 3, which means we need to integrate three areas a^2 to get one perfect cube
Please pay good attention to this basic geometry as it has lot of relevance to what to whatever I discuss in this book

Surface Area of a cube:
Surface Area of a cube of side a = 6 a^2

Volume of Cuboid:
Volume of Cuboid = height x width x length, if we consider a Cuboid of height a, width a and length 2a then Volume of Cuboid = a x a x 2a = 2a^3

Point: Point is pi for the geometry of sphere and 1 for geometry of cube

To sum up if we keep on integrating, geometry of circle starts with a unit of pi which becomes a circumference of circle , which becomes area of circle, which becomes surface area of sphere and which finally becomes volume of sphere and this process repeats all over from fourth dimension

Similarly if we keep integrating, the geometry of cube starts with 1 becomes a line, which becomes circumference of a square, which becomes a square, which becomes surface area of a cube and which finally becomes the volume of cube and this process repeats from fourth dimension

If we put the consecutive components of sphere in a series we get pi, pi*r, 2*pi*r, pi*r^2, 4*pi*r^2, 4/3*pi*r^3, 2 * 4/3*pi*r^3

Similarly if we put the consecutive components of a cube in a series, we get 1, a, 2a, a^2, 3a^2, a^3, 2 x a^3

Chapter 2
Theory of Reigning Element

Theory of Reigning Element

Or

Principia Gravitas

Or

The Grand Design of Universe

Or

Theory of Everything

Or

Universe in a Snail Shell

Every particle has a reigning element (which shall be hitherto referred to as R.E.) which determines its gravity, the gravity of the particle is the residual super mass left out after forming the most stable element (which shall be hitherto referred to as S.E.) for the given reference frame (which shall be hitherto referred to as R.F. and will be explained later), no gravity can be greater than 120 in the same R.F..

Here I need to elucidate the term R.F., an R.F. is a reference frame represented by the third dimension of the pure energy vector Ve given by Ve = (C, C^2, C^3, C^4, C^5, C^6) where C is the speed of light for the given R.F., C is constant for a given R.F(equal to 3x10^8)generally defined as the number of parts into which a Photon from higher energy dimension divides when it enters a lower energy dimension is the speed of light but differs in relation to other frames, thus

C01 w.r.t. R.F.3 = 1/(27x10^64)

C21 w.r.t. R.F.3 = 1/(3 x10^8)

C32 w.r.t. R.F.3 = 1x10

C43 w.r.t. R.F.3 = 3x10^8

C54 w.r.t. R.F.3 = 27x10^64 and so on for R.F. -1, 0, 1, 2, 3, and 4 respectively as compared to our R.F.. and C^4 is gravity, C^5 is gravitophoton and C^6 is the source of energy(which has shape of a perfect cube and originates at Sun)

However mass vector has five dimensions given by vector Vm = (x, y, z, g, P) where x, y, z are three physical dimensions (magnitude of these first three common dimensions of mass and energy vectors given by (ct)^2 and has a shape of cylinder), g is the graviton, p is the prasadon(HIgg's boson) and magnitude of the mass vector is given by cube root of ((ct)^2 + g^3 + p^3)^1/3 and has a shape of sphere

So there are 11 total dimensions(5 mass and 6 energy) they club together to form 3(P,p,e) mass dimensions including one common physical dimension(e) having three dimensions and three energy dimensions making the total five dimensions

GRAVITOSTATICS OF STABLE SYSTEMS:

We know that Einstein's equations do not cover the quantum mechanics, to include the quantum mechanics though unobservable is very much a part of the reality we need to combine the two systems

Stable system S comprises of three co-ordinates C^3, g,P in which the equations of Newtonian mechanics hold good and in third coordinate t Einstein's relativity holds good which in turn is one more three dimensional system relative system R having three real physical coordinates x, y, z the magnitude of which is given by $(x^2 + y^2 + z^2)^{1/2} = (ct)^2$.

If we wish to combine the two systems as in the equation given below

$$p^3 + g^3 + (ct)^2 = (CT)^3, \text{ where } C = c^3 \text{ and } T = t^2$$

or if we consider the three physical coordinates x, y, z

$$p^3 + g^3 + x^2 + y^2 + z^2 = (CT)^3$$

This equation can be envisaged as a rigid sphere of radius (CT) with its centre at the origin of co-ordinates of stable system S with two static coordinates of P, g and moving third coordinate $(ct)^2$

The transformations are given by slightly modifying Lorentz transformations as given below

```
p'  B^3  0    0   0  0  p
g'  0    Y   -YB  0  0  g
x'  0   -YB   Y   0  0  x
y'  0    0    0   1  0  y
z'  0    0    0   0  1  z
```

In other words, sum of first five natural numbers is mass(5 x6/2 =15), sum of squares of first five natural numbers is energy (5 x 6 x 11/6 = 5 x 11) and sum of cubes of first five natural numbers is super mass or gravity (15^2)

In a nutshell the pure energy which is in the form of perfect cube transforms to the form of a perfect sphere of mass by passing through the transition phase of electrodynamics which is in the form of a cylinder

Element with Gravity 119 is Photlon (containing 119 super protons) of next R.F and cannot be classified as Element

Element with Gravity 120 is super Neutron(containing 120 super protons) of next R.F. and also cannot be classified as Element

Element 121 which is super Hydrogen can be aptly named Gravitlon and constitutes the super mass.

As the Gravity of any Element for a given R.F. cannot exceed +120 our present periodic table ends at 120 and the same elements with higher energy dimension will fill the next periodic table starting from 121 ending with 238 i.e. 121 will be the super Hydrogen 122 the Super Helium and so on.

Energy to Mass Conversion

Sun is the super beryllium or element 124 each particle of which can be called SunIon has energy 10^46 joules, which divides into three parts namely element 120 or Magneton with 10^6 photons, element 121 or Graviton with 10^16 Photons and element 122 or PhotIon with 10^24 photons which divides into cube of ThermIons which contains (10^2)^3 = 10^12 photons, each of which further divide into cube of MagnetIon which contains (10^2)^3= 10^8 of photons, each of which divides into cube of ten Sonions which contain ten photons, each of which ultimately divides into one Proton.

Proof for this can be found in spectrum of Sun (I would rather call it Gravito Photo Thermo Magneto Sono Protonic spectrum or GLTMSP spectrum).

So pure energy divides into various forms. Lumion is light, ThermIon is heat, MagnetIon is Magnetism, SonIon is sound, 10^8 and 10^2 are right angles to each other, 10^4 and 10^2, 10 and 1 are at right angles to each other, 10^4 is radiated as heat energy otherwise known as Hawking radiation, this radiation of free ThermIons (perhaps) coalesces to form the E(u)ther(mions) which pervades the whole empty space and supplies the deficient energy needed for Sun's fission.

Thus square of photon is sound, square of sound is magnetism, square of magnetism is heat, square of heat is light, square of light is gravity

Thus all forms of energy belonging to same R.F. are inter convertible (Specific rules apply which will be explained later) and Energy is limited and conserved in any given R.F..

One Joule increase in Thermal energy causes the temperature to raise by 1 Degree Celsius.

When Sonion ultimately divides (w.r.t.R.F.3) into individual proton the wave function collapses as proton is the basis for our reference.

When heat is increased the magnitude changes indicated by rise in temperature, i.e. if heat increases by 10^4 joules there will be a corresponding rise of one degree in temperature(the ThermIons become the ThermIons of next element, i.e. if we raise the temperature by one degree ThermIons of hydrogen will form, if we raise by 2 degrees helium and so on until 120 thus 120 degrees Celsius is maximum we can raise the temperature of element 120 or 144000 degrees is maximum temperature achievable in a given reference frame thus the entropy of a inertial reference frame is limited and conserved).

This explains the anomalous expansion of water which when heated contracts till 4 degrees and then expands as ions till Beryllium ions still constitute the mass or atomic world.

So when 119 packets of any of these fundamental forms of energy come together they gain one dimension and when four such packets come together then they become one bigger Ion of energy, each of which is known as one Mole, thus each of these packets contains Avogadro's number i.e. 6 x 10^23 of particles, this is also the exact reason why ground water starts evaporating at 37 Degrees Celsius, ground water has 82 Thermlons and an increase of 37 more makes it 119 at which it gains higher fourth dimension so converts into vapor

In other words Super Mass is Gravity.

Earth has element 92 as R.E., i.e. Uranium is the highest possible naturally occurring element on planet Earth, so 92—82 = 10 is its Gravity.

Energy behaves as particle when it is exact multiple of 10^8, 10^4 and 10^2 units of energy and as wave when it is fraction of the same.

Energy occupies maximum space which is the cubical form and is represented by exponentials of 10 while mass occupies least space which is spherical form and is presented by exponentials of 11

So when Photons from other sources go nearby Sun they bend as there is a slight warping of space caused by the difference in charges 119 and 124 but inside solar system the warping is total (the system from outside looks like a closed sphere) in the sense the light from this system does not go out but only that of Sun would be seen from higher energy dimensions, in other words Sun is holistic representation of our system and thus the fourth dimension for our world.

Thus every second of a second we are observing a big bang(fission of 10^46) and a big crunch(formation of Beryllium atom), in other words singularity of one R.F. is big could cause big bang of lower R.F..

My dear cute big Genius with beautiful mind Dr. Hawking, I proved you right.

This was the story behind the successful cold fusion achieved sometime back in 1989 using beryllium atoms by Martin Fleischmann and Stanley Pons .

The time arrow always points to the Past unless someone can actually stop the great wheel of time and force big bang from big crunch .and there by reverse the direction of time

So my dear brother Dr. Mallet, I am afraid journey into past is ruled out unless you have nexus with God (who alone probably can have the strength to push things back to Big bang from Big Crunch), however travel to future is quite possible as we may very likely create energy equal to or more than C^2 (light particle is reported to have been teleported).

The gravity for the planets of our solar system aregiven in the table below.

The gravity for the planets of our solar system are given in the table below.

Planet	Reigning Element	Gravity	Formula	Orbital
Mercury	84	2	1(3E.D.)^2+1	S
Venus	88	6	2(3E.D.)^2+2	P
Earth	92	10	3(3E.D.)^2+1	D
Mars	100	18	4(3E.D.)^2+2	F
Jupiter	108	26	5(3E.D.)^2+1	G
Saturn	120	38	6(3E.D.)^2+2	H
Uranus	121	2	1(4E.D.)^2+1	K
Neptune	122	6	2(4E.D.)^2+2	L
Pluto	123	10	3(4E.D.)^2+1	M
Total Gravity		118		
Sun	124	124		

Note: E.D. = Energy Dimension

Note: E.D. = Energy Dimension

Now let us start with pure energy vector and try to arrive at the mass vector Please refer to the illustrations given below:

Yes, we can unify Quantum Mechanics and Relativity

Hypothetical Union of the Ions of Energy

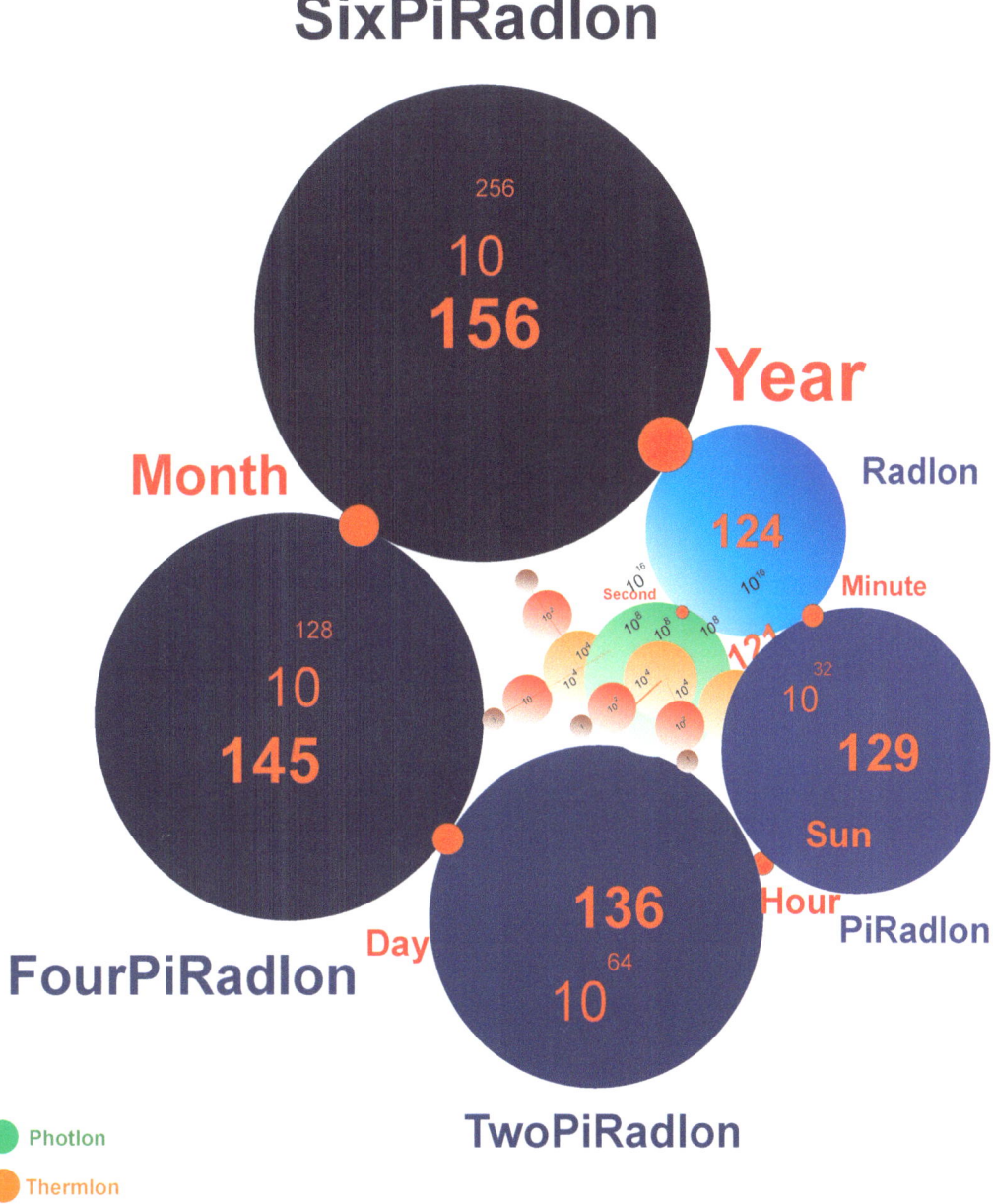

IdealUnIon

Prasad Vemulapalli

illustration Ideal Union shows the union of various forms of energy immediately after Big Bang when Energy was in its purest form and no matter has yet formed, pure energy spreads in spherical waves and occupies maximum space. We can see that C^3 is a second C^6 a minute, C^9 an hour, C^{12} a day, C^{15} a year, so as Sun has energy(including cosmological constant of 10^2) of is $10^{48} = (10^6)^8$ it takes 8 minutes for Sun light to reach Earth. However after the big bang there has been much interaction so now the union of ions will look as shown in illustration Energy Real(Un)ity

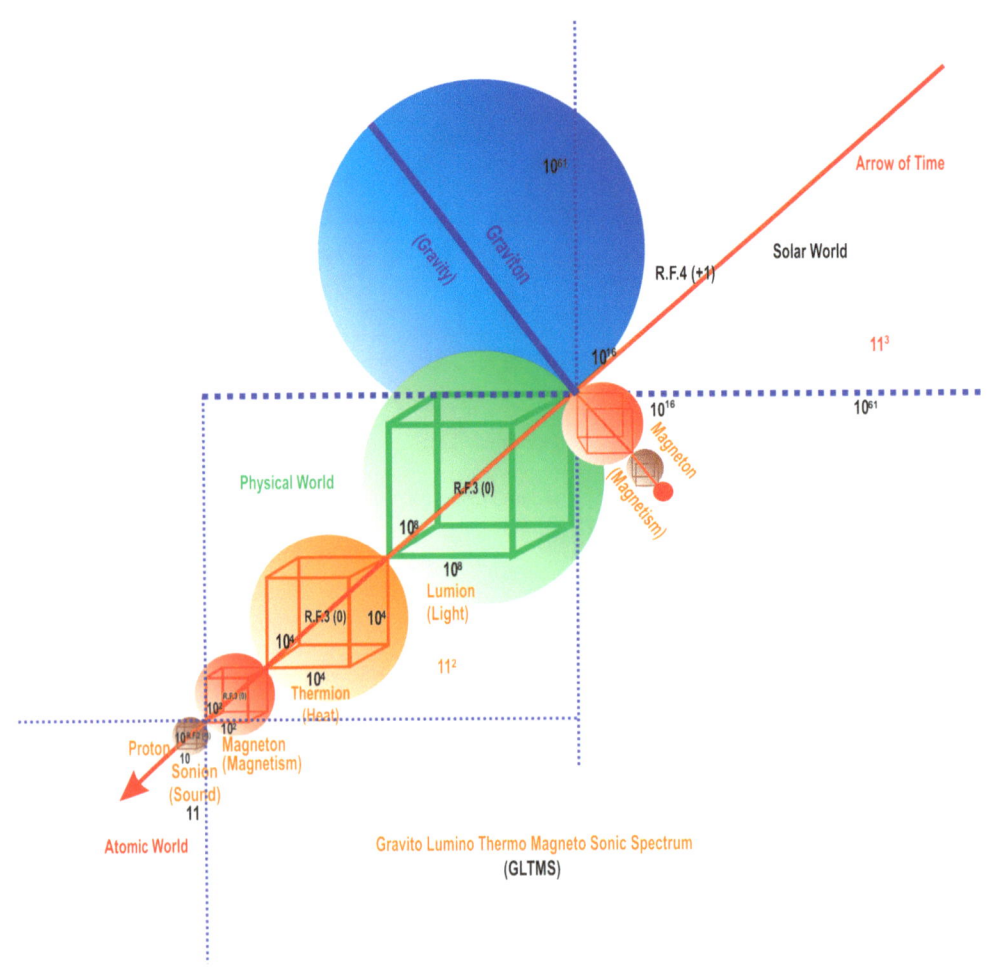

When the maximum volume is reached the energy ions started cooling down and matter started to form, at this stage the ions of energy of same type meet each other and form cubes of energy which look as shown in illustration UnIon

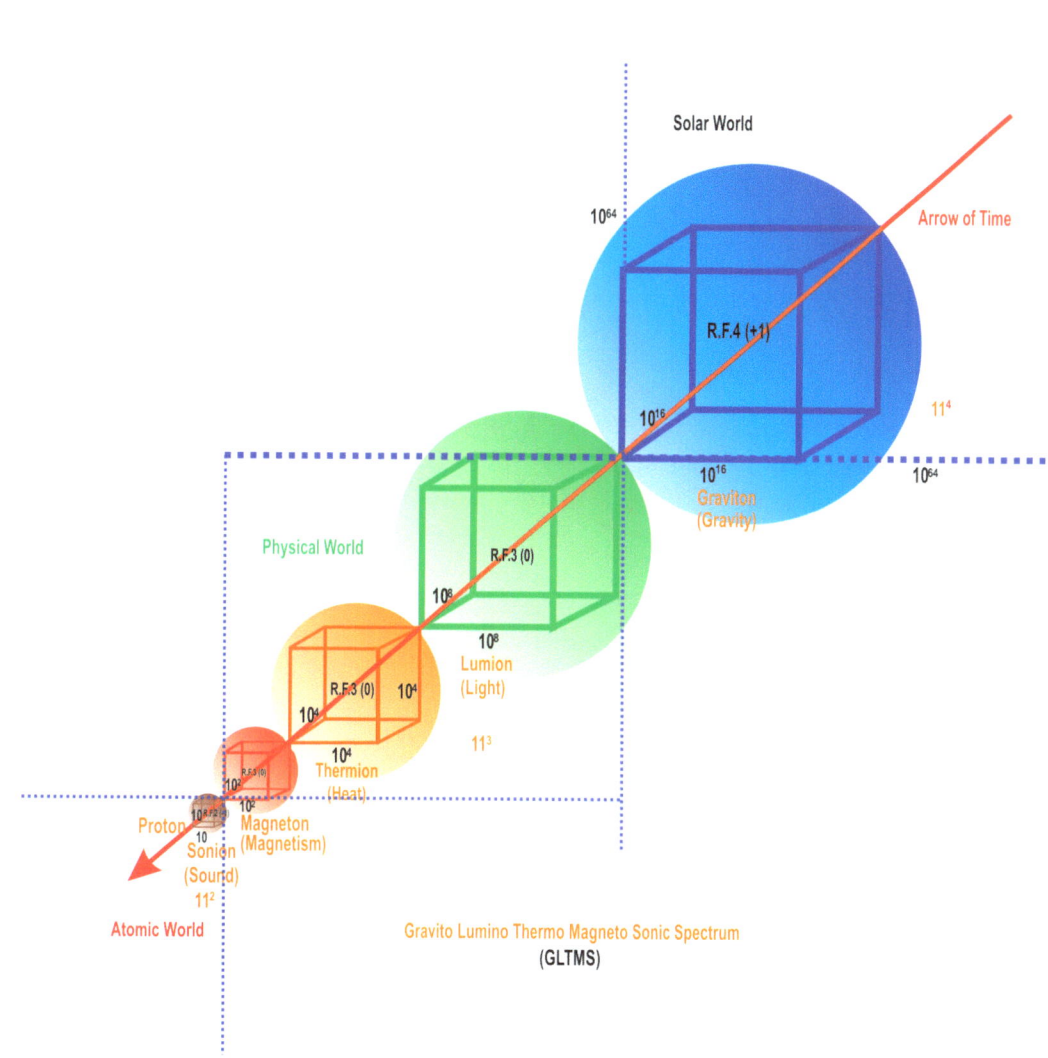

If we consider one single time lines

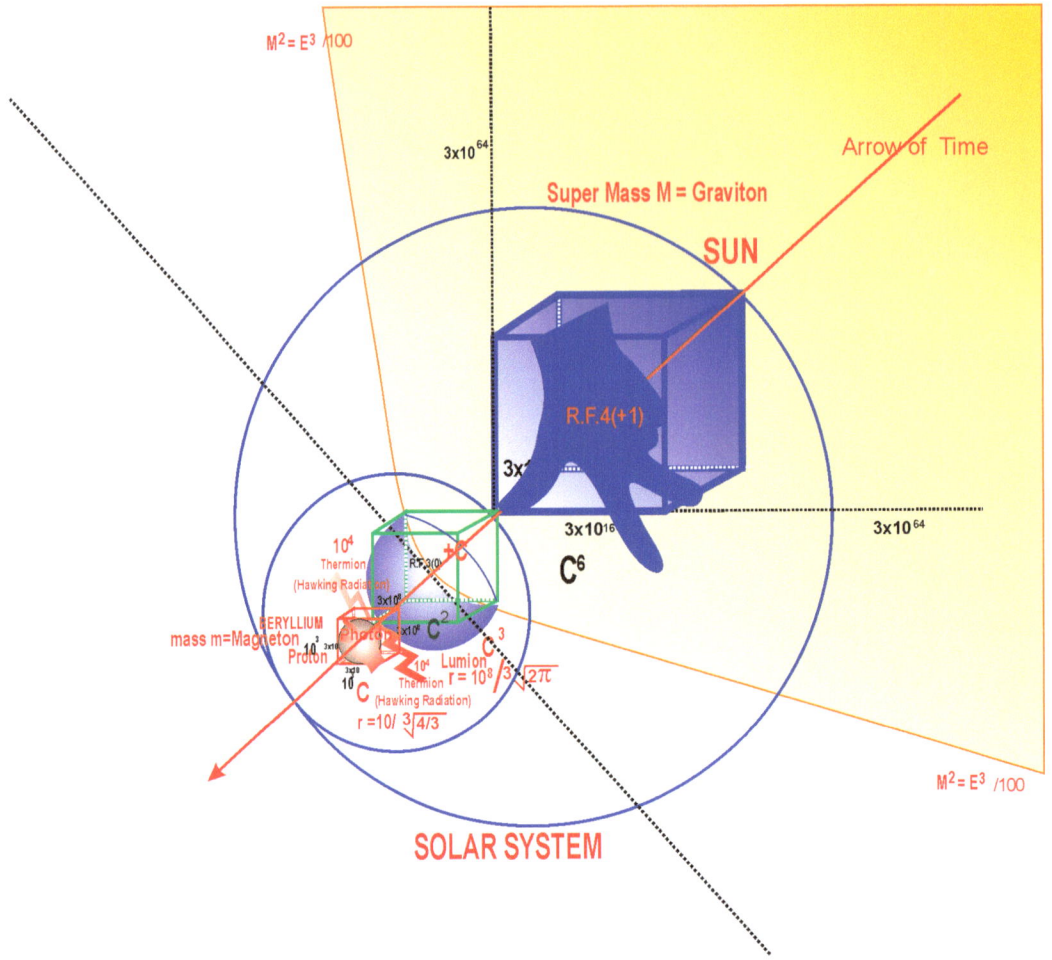

Real(Un)ity

Mass occupies least space so takes spherical form, Proton is the point, Hydrogen is Arc of 90 degrees, Helium is arc of 180 degrees(or pi), Beryllium is circumference of a circle, Planets are area of circle, Energy of planet is surface area of the sphere and Sun is the sphere

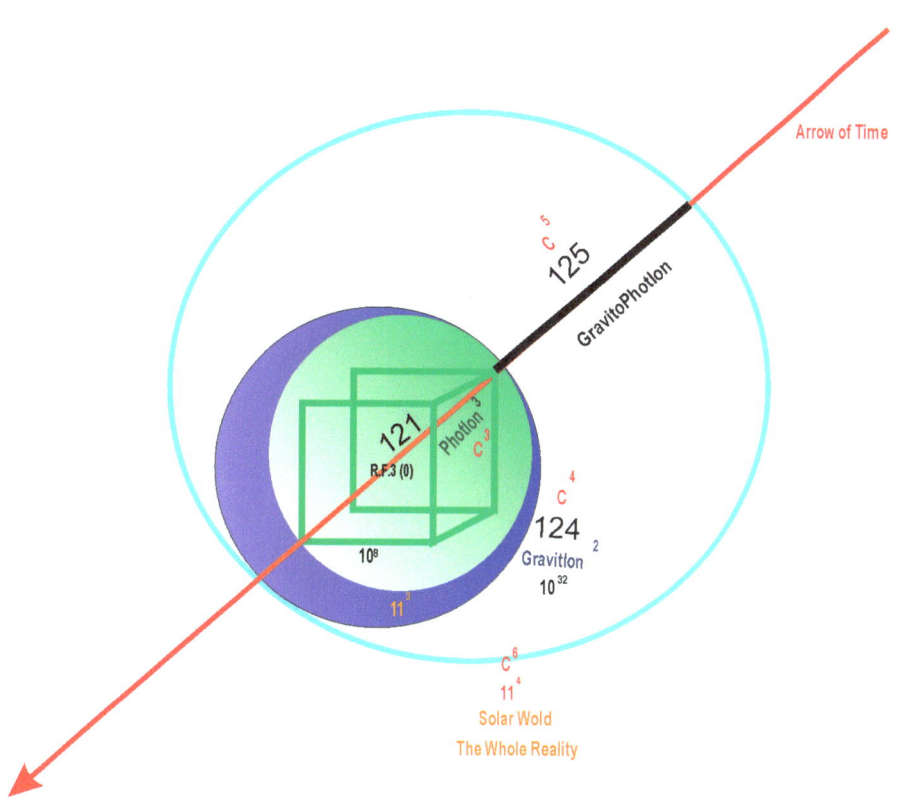

Gravity is square of Gravitlon and has four Cs, which form the hollow sphere (4pir^2) of Gravity

Prasad Vemulapalli

Reality for the holistic Solar System.

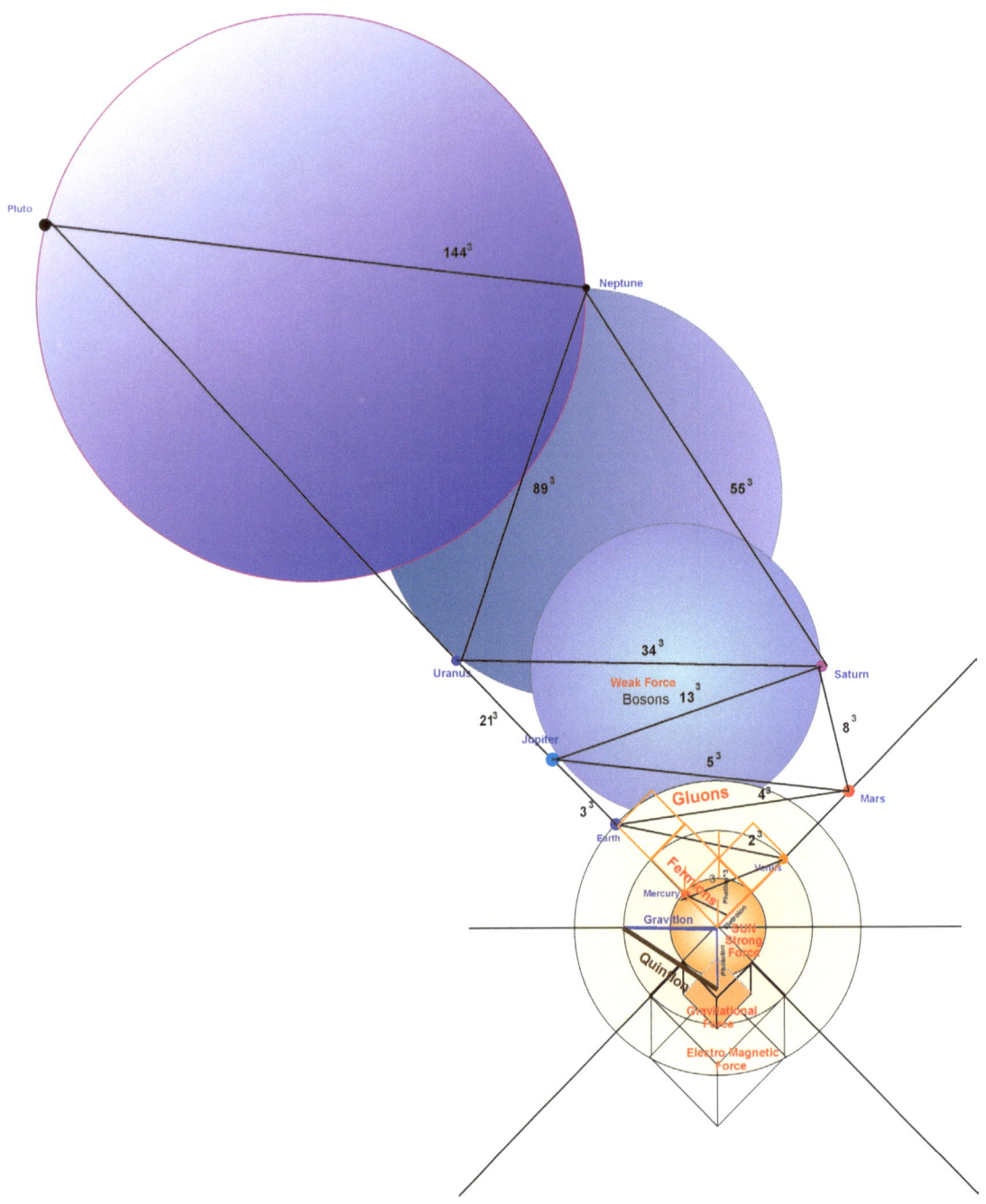

SunReal(Un)ity

Yes, we can unify Quantum Mechanics and Relativity

Einstein proposed that if you increase the velocity of a given mass to the velocity of light then the mass increases infinitely but he did not tell us what this infinite mass could be, in fact when given the velocity of light, it becomes a Photon whose rest mass is zero and when given square of velocity of light becomes the super mass and gains one energy dimension, so after all the infinity of one dimension is zero of next dimension, i.e. the threshold of the next dimension.

According to theory of relativity $E = MC^2$, so if the particle has velocity C then $E/C = MC^2/C = MC$ which is super mass with ve¬locity of light and thus nothing but Photon with zero mass on the other hand if particle has velocity C^2 then $E/C^2 = MC^2/C^2 = M$, which is nothing but super mass

On earth the mass of proton is 1.67×10^{-27} kg, so one kilogram contains $(1/1.67) \times 10^{27}$ protons $= 6 \times 10^{26}$ = weight of earth) $\times 10^2$, so mass of any particle is the number of protons it receives per second from the source of energy $\times 10^2$(this being the mass or energy of each of the dark particle and these were the Helumes I saw in 1992) or mass can be in other words called energy frequency of a particle.

If we just redefine Energy in terms of Frequency, we can unify Relativity and Quantum theories

Reference Frame	Energy Constant	Mass Density	Frequency	Wavelength	Frequency x Wavelength
Atomic World or Pure Mass	Planck's	$119 \times 1.67 \times 10^{-27}/3 \times 10^8 = 6.625 \times 10^{-34}$	1 to 3×10^8	3×10^8 to 1	C
Physical World or Mass and Energy	Newton's	$2/3 \times 10^8 \times 10^2 = 6.67 \times 10^{-11}$	3×10^8 to 9×10^{16}	9×10^{16} to 3×10^8	C^3
Solar World or Pure Energy	Einstein's	$3^2 \times 10^{16} \times 10^2 = 9 \times 10^{18}$	9×10^{16} to 27×10^{24}	27×10^{24} to 9×10^{16}	C^6
Extra Solar World	Prasad's	2×10^{48}	27×10^{24} to 81×10^{32}	81×10^{32} to 27×10^{24}	C^9

According to relativity, Energy is given by $E = MC^2$

According to Theory of Reigning Element Energy is given by $E = mv$, where m is mass potential for the given R.F. (which is mass of proton for atomic world, mass of super proton for physical world and mass of Sun for solar world) and v is mass frequency which is the number of particles the particle from higher region divides into when it enters lower energy dimension.

We can thus combine quantum theory and relativity.(since we are defining energy of fourth dimension also in terms of discrete particles having well defined frequency)

In other words we can generalize the energy E = mC for atomic world, (m+M)C^2 for physical world and MC^2 for solar world

Please keep in mind that energy always takes form of cubes which occupy the highest space for a given unit of length and mass always takes form of sphere which occupies least space for a given unit of length

All the hypothetical structure are shown in the form of cubes and the reality is shown in the form of spheres or sub divisions of the same such as cones or cylinders (three cones make one cylinder)

Time as we all know has three components Future, Present and Past.

Future is pure kinetic energy in the form of square of Gravitlon C^4, which divides into Present C^3(which is perfect cube which divides into one Gravitlon C^2, the potential energy represented by 2/3 sphere and 1/3 kinetic energy C represented by 1/3 cube) and Past which is pure potential energy mass in the form of perfect sphere and has potential energy is C

Future is real Time (denoted by +1) and has energy C^4. Present is the event horizon(denoted by 0) and has energy C^3. Past (denoted by -1) is the imaginary Time and has energy C,

Thus for any event Future + iPast = Present, where i = -1^1/2.(C^4 x C^-1 = C^3)

If we define time in terms of energy it has three components namely Magnetic, Photo Electric and Gravitational components, and it is the cross product of all three components in other words it is cross product of the Potential Energy caused by magnetic component, Kinetic Energy of Photo Electric component and Gravitational Potential Energy caused by Gravity

Thus if we assume the angle between each to be 90 Degrees, we get

Energy E = Magneton x Gravitlon x Photlon (1),

Substituting values for Magneton, Photlon and Gravitlon,

We get Total Energy = C^6C^24*C^16 = 10^46

According to relativity, Energy of Sun = 2x10^30x9x10^16 = 18x10^46 joules.

In other words speed of light C is the yardstick for energy dimensions, i.e. C^0 =1 is the zero'th dimension, C is first dimension, C^2 is second dimension, C^3 is the third dimension and C^4 is the fourth dimension and so on.

So this is simply rephrasing of relativity.

The energy of three individual Gravitlons is $G^3 = (10^{16})^3 = 10^{48}$ joules but Sun has only 18×10^{46} joules , so the difference of $(2 \times 10^{16} + 10^{14}/10^{48})$ = 0.01 between the mass of individual components (three Gravitlons) and the combined holistic particle of sun accounts for the mass defect. The difference factor of 10^2 is cosmological constant which accounts for the zero point energy of space. Einstein was not wrong in introducing the cosmological constant to make up for the missing energy. As we can see from illustration the hubble's constant is the vector difference of powers of positive velocity of light and the negative velocity of mass or the mechanical mass wave, i.e. $10^{-81} \times 10^8 \times 10^2 = 10^{-71}$, the red shift is in fact the acceleration of the mass in the opposite direction to the source of energy.

So practically the total energy of Sun is Energy $E4 = 1 \times 10^{46} = C^6/10^2$ = $G^3/100$ for one particle so it can be formulated as $E4 = M4 \times G^3 /10^2$ for particle of super mass M

Practically it is suffice to know the three Energy Constants, namely Planck's Constant(6.625×10^{-34}) for R.F.2 or Atomic world, Newton's Gravitational Constant(6.67×10^{-11}) for R.F.3 and Einstein's Constant for solar world (9×10^{18}) for R.F.4 to calculate any energy of these three worlds

As mentioned already all these four entities namely Gravitlon, Photlon, Thermlon and Magnetlon are inter convertible, they are like currencies of each of the four dimensions of same R.F.,

Energy of expressed in terms of pure Gravitlon or pure Photlon or pure Thermlon or pure Magnetlon is given below:.

Prasad Vemulapalli

According to Theory of Reigning Element., Energy of sun in terms of

I) **Pure Graviton** is given by Time = 2 x 9 x (10^16)^2 x 10^8 x 10^4 x 10^2= 18x10^46= Ee x C^7/4, where Ee = 2 x 9 x 10^32

II) **Pure Photon** is given by Time = 2/(3 x (10^4)^2 x 10^2) x 27 x 10^24 x 10^16 x 10^8 x 10^4 x 10^2 x 10^2 = 18x10^46= En x 10^56, where En = 2 x 9 x 10^-10 (the magneton and the cosmological constant offset each other)

III) **Pure Thermion** is given by Time = 2/((10^16)^2 x 9 x 10^8) x 81 x 10^32 x 10^24 x x10^16 x 10^8 x 10^4 x 10^2 = 18 x10^46.= E p x 10^ 88, where Ep = 2 x 9 x10^-40

IV) **Magneton** is given by Time = 2/(81 x (10^32)^2 x 10^16) x 729 x 10^48 x 10^32 x 10^24 x10^16 x 10^8 x 10^4 x 10^2 x 10^2 = 18 x10^46.

We can observe some interesting developments when we try to calculate Energy in terms of pure Sonions and Protons

I) **Pure Sonions** is given by Time = 2/((10^64)^2 x 729 x 10^32) x 2187 x 10^64 x 10^48 x 10^32 x 10^24 x x10^16 x 10^8 x 10^4 x 10^2 x 10^8 = 18 x10^46.

II) **Protons** is given by Time = 2/(2187 x (10^128)^2 x 10^64) x 6561 x 10^96 X 10^64 X 10^48 x 10^32 x 10^24 x10^16 x 10^8 x 10^4 x 10^2 x 10^2 X 10^72= 18 x10^46.

If we observe carefully, while calculating the Sun's energy using the pure Gravitons and Thermions the zero mass energy was omitted because these two energies are self sufficient and pure energies and do not contain any mass .

Universal Gravitational Constant G given by Newton is combination of Photions and Magnetion =2/(3 x 10^8 x 10^2) or 2/(C x 100) = 6.67 x 10^-11 which means Newton considered only the Photions and Magnetions for calculating the gravity which works perfectly fine for velocities below C but once they approach C it does not work anymore on the other hand the super gravity which uses Gravitions works for all worlds.

Thus we confused gravity with super gravity all the time and were wondering why it is not working for high speed(equal or more than C) particles.

Even we can calculate the energy of Sun in purely Newtonian way using universal gravitational constant as 2/(3x10^8x10^2) x 27x10^24x10^16x10^8x10^4x10^2x10^2=18x10^46.

Element 82 i.e. lead is the S.E. for our R.F. having coordinates(x, y, z, 3), similarly the element 122 is the S.E. for next R.F. that of the family of sun having coordinates(x, y, z, 4).

All sources of free unbound super energy will radiate energy in packets given by the relation E= MxC^2 until they become stable, for e.g. our sun will radiate energy in packets of 9x10^16 joules and eventually becomes element 122(white dwarf) or 120(neutron star) .

Here what I have illustrated in Uni(I)on is purely hypothetical but in reality the situation is different as illustrated in Real(un)ity, atomic world is fully evolved and appears as holistic beryllium atom and is completely mass w.r.t. to our world represented by perfect sphere and accounts for magnetism, which can be aptly called Magnetlon, our world is partially evolved so it is 2/3 mass represented as sphere and 1/3 energy represented as cube, so we have both 2/3 electricity and 1/3 gravity in our world and Solar world is still of course purely energy so represented by perfect cube and is pure gravity.

This reality however is almost valid for a planet mercury which has 2/3 mass and 1/3 energy but for others the crescent portion of mass goes on increasing till it becomes perfect sphere for planets starting from Saturn as they belong to 4 E.D. and they receive only the Photlons and no Gravitlons at all .

However Saturn is little different for the reason that its gravity is 120 which makes it neutral and stable that it is neither 3 E.D. nor 4 E.D. so it is torn between the two worlds and which explains the rings .

The 2/3 rule is in the very nature of things, even human beings can at the most carry a weight which is 2/3 their weight so do the planets(which belong to 3 E.D.) so they can have satellites sum of whose gravity is 2/3 of their own gravity.

To host life we need a planet having gravity of 92(82+10) as water has gravity 10(H20= 2x1+8=10)) , as planets starting from Saturn belong to energy dimension of the sun they have gravity more than 120, so we can try for suitable satellite of one of these planets, first of which could be Titan, if it is not possible in our solar system, then we may have to search for such a planet in neighboring systems .

A stable system can have any number of planets whose cumulative gravity does not exceed 118.

R.F.s are consecutive and relative to each other, thus our R.F has third energy dimension, atomic world has two energy dimension and solar world belongs to fourth energy dimension.

Otherwise if we consider our world as inertial reference frame then we can consider our world as zero, atomic world as -1 and solar world as +1.

Each of these reference frames are rigid, fixed to the next reference frame so are inertial reference frames in relation to higher reference frame

Thus all concepts of quantum physics and uncertainty principle hold good for the atomic world Newton's laws hold good for our physical world and relativity holds good for the solar world. These worlds are all real in their own time but seem to be imaginary(past) and illusionary(Future) in relation to each other, if only one has precise means to measure the fraction(for atomic world and below) and multiple(for solar world and above) of speed of light one can calculate energies absolutely and then **there will be no uncertainty**

Balance Sheet of Sun's Energy.

R.E. Energy = 9*10^18*Energy Frequency

Balance Sheet of Sun's Energy.
R.E. Energy = 9*10^18*Energy Frequency

Planet	Orbital	Gravity	Mass	Energy Frequency	Velocity in m/s	RE Energy	GR Energy
Mercury	S	2	3.30E+23	3.30E+21	446387.639	2.97E+40	2.97E+40
Venus	P	6	4.87E+24	4.87E+22	326110.198	4.38E+41	4.38E+41
Earth	D	10	5.98E+24	5.98E+22	277777	5.38E+41	5.38E+41
Mars	F	14	6.50E+23	6.50E+21	222777.154	5.85E+40	5.85E+40
Asteroid Belt		4	-	-	-	-	-
Jupiter	G	26	1.90E+27	1.90E+25	120555.218	1.71E+44	1.71E+44
Saturn	H	38	5.70E+26	5.70E+24	89721.971	5.13E+43	5.13E+43
Uranus	K	2	8.70E+25	8.70E+23	63333.156	7.83E+42	7.83E+42
Neptune	L	6	1.00E+26	1.00E+24	50555.414	9.00E+42	9.00E+42
Pluto	M	10	7.00E+23	7.00E+21	44166.543	6.30E+40	6.30E+40
Sundry*		6	3.05E+25	3.05E+23		1.98E+46	1.98E+46
Total		118	2.70E+27	2.70E+25		2.00E+46	2.00E+46
Sun		124	2.00E+46	2.70E+25	C^2	2.00E+46	2.00E+46

Yes, we can unify Quantum Mechanics and Relativity

To sum up the observations, if we compare the atomic world and solar world we find the equivalence between the various fundamental particles.

Sun is the nucleus made up of one +1/2 upward quark(PhotIon having 10^24 Photons), one +2/3 quark(GravitIon having 10^16 Photons) and one -1/2 Lepton having -10^8 Photons)

Saturn is the Neutron

Planets below Saturn are Mesons, made up of PhotIons, ThermIons and MagnetIons, which together with GravitIons constitute Hadrons

Neptune, Pluto are the W &Z Bosons

In other words Planets below Saturn are Electrons, their satellites are Muons and the Asteroids are the Tau, all of which together form the Leptons

Tachyons are lower forms of life with four dimensions

The C-Symmetry, P-Symmetry and T-Symmetry as you can observe from the illustration S(un)Real(Un)ity are all obeyed and violated as per the established science

I would appreciate little help here in mapping the planets to the elementary particles (As I mentioned earlier my knowledge of QM is very limited)

Chapter 3
Mass to Energy

In Chapter 2 we have seen that each reference frame has limited energy given by -C, C^3, C^6, C^9 and so on represented by 11^2, 11^3, 11^4 and so on and as we can observe from illustration Real(Un)ity 11^2 = 10^3, 11^3 = 10^24 and 11^4 = 10^64

Let us now see what we get if we extend the series below 11^2, logically we should get 11^1 = 10^1, which means mass becomes energy so proton becomes photon for the sub sub atomic world

So let us slide down the window of Reality one step and peep into atomic world and see how the mass converts into energy

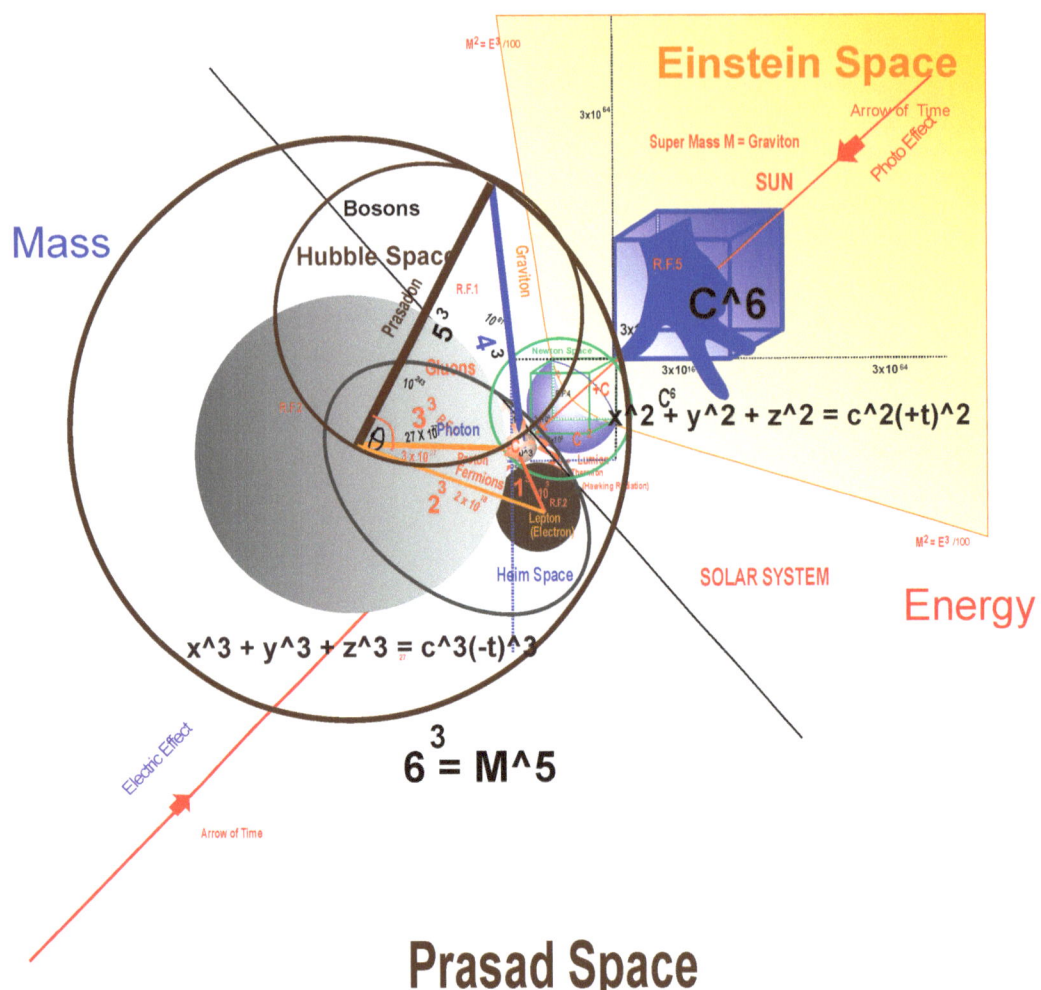

Prasad Space

10^2 is the least unit on energy side and 10^3 is the least unit of mass on mass side of the spectrum, so 10 being the speed of light for the atomic world, we can conclude that C^2 is energy and C^3 is mass for any frame(cross product of potential and kinetic parts of energy)

Coming to Mass-Energy equivalence, the base relation is (C^3)^2 = (C^2)^3 = C^6 or M^2 = E^3 but in practice we observe that the particle of Sun has only 10^46 not 01^48 so there is something missing of the order of 10^2 which is the cosmological constant which the sun gets from the dark matter, this is the reason why energy is released when ever the magic figure of 100 appears as in (1 + 2 + 3 +4)^2(cold fusion, BTW this is the string with 10 dimensions that string theorists believe in, so string is composed of 10^2 photons or 10 protons) or 6^2 + 8^2 (chemical reaction as in carbon(which is in mass form so already squared) + oxygen(which is gas so treated as ion) or 10^2(Nuclear Fission) or i^3 + 2^3 + 3^3 +4^3(Nuclear Fusion or super string on solar scale) , we can also express the same in other ways as explained below, we know that a mole contains Avagadro number of particles i.e. 6 x 10^23, so (6 x 10^23)^2 = 2 x 18 x (10^16)^3/ x 100 or 18 x (10^24)^2 / 3 x 100 which means a mole consists of either two cubes of gravitons or two squares of PhotIons or particles of Sun make one mole Mole^2 = G^3/3 x 100 or 2 x P^2 where G is Graviton and P is the PhotIon

Currency Exchange between Mass and Energy:

If we look at the Whole Real(Un)ion illustration we observe that 15^2 = 1*1^2 + 2*2^2 + 3*3^2 + 4*4^2 + 5*5^2 = 6^3 + 3^2, there are two important consequences of this identity, first the definition of fine-structure constant and second the exchange rate of currency between the quantum world and physical world

The fine-structure constant:

Now let us try to understand the language of quantum world ,we have already seen that the least unit of mass is 10^3 so every quantity of physical world is represented by the reciprocal of the power of 10^3, 10^3 is represented as 1/3^1,10^6 as 1/3^2 and so on, so 10^18 is represented as 1/3^6 which is superfine structure of quantum world and since the energy for quantum world comes in packets of 10^6 (magnetions), fine structure constant is expressed as 10^6/3^6 or 1000000/729 = 137

Simplifying In other way we can say the quantum numbers are the reciprocal of logarithems to the base 10^3 of our world , since the quantum world is upside down, and if we are in quantum world then we can as well simplify represent them as simply the logarithems of 10^3 and I am very sure that all fundamental particles are exact multiples of 1/10^3, I hope I would probably be able to provide the formulae someday(simpler than Heim's)

Rate of exchange of currency between quantum and physical worlds:

If we observe the identity 15^2 = 6^3 + 3^2, it is clear that there is a deficiency of 3^2 in fine-structure to fill in the 15^2 which is the sum of squares of natural numbers till 5. If we observe the illustration Energy Real(Un)ity, on the energy side of the picture we have a clear deficiency of 10^2(Sun's energy is just 18 x 10^46 while we need (10^16)^3 or 10^48 to get energy), so mass attains perfection by accepting energy in packets of 10^6 while energy does so by accepting 10^2 or (10^6)^1/3 so MagnetIon is the currency that is accepted in quantum world while Magneton is accepted in the physical world, so whenever mass is converted to energy we have energy relased in multiples of 10^2 and whenever energy is converted to mass It is consumed in multiples of 10^6

Prasad Vemulapalli

Our Solar system can be visualized as the vortex formed by following the triplet sets of pythogorean cubes 3^3 + 4^3 +5^3 = 6^3

Definition of pi:

Since there are total of 11 dimensions forming 5 actual dimensions in each elementary particle (if we consider Sunlon as one), if two such particles come together they form 2 x 11 = 22 total and 10 actual dimensions, of the 10 actual dimensions three lower dimensions merge together forming the core leaving 7 actual dimensions, thus pi has 22 total dimensions and 7 actual dimensions giving its value 22/7

Mass Energy Equivalence

Mass:

1 = (spin 1)

1 = 1^2(-1/3 down quark)

1^3 = 1(Electron)

1 + 1 = 2(spin 2)

2 X 2 = 2^2(2/3 up quark)

2 x 2^2 = 2^3(Magneton)

1 + 2 = 3(spin 0)

1^3 + 2^3 = 3^2(gluon)

3 x 3^2 = 3^3(Proton)

3^2 + 4^2 = 5^2

5 x 5^2 = 5^3(Prasadon)

4 x 3^2 = 6^2

1^3 + 2^3 + 3^3 = 6^2(Heim's Metron)

6 x 6^2 = 6^3

3^3 + 4^3 + 5^3 = 6^3(Prasad's Space)

Mass can be created in any combination of the above particles

Energy:

1^3 + 2^3 + 3^3 + 4^3 = 100(Cold Fusion)

6^2 + 8^2 = 100(Nuclear Fission)

10^2 = 100 (Nuclear Fusion)

Energy can be created in any of the above combinations

As we know $M^2 = E^3$, the general equation for Mass-Energy Equivalence can be summarized in one equation as $x^3 + y^3 + z^3 = c^3(-t)^3$ or for general solution

$(((n + 2)^3 + (n+3)^3 + (n + 4)^3)^{1/3})^2 = ((n^2 + (n +1)^2 + (n + 2)^{1/2}))^3 /100$

Or raising both sides to power of 6

$((n + 2)^3 + (n+3)^3 + (n + 4)^3)^4 = (n^2 + (n +1)^2 + (n + 2))^9 / 100$

The integral solution to this equation provides the various particles

We know that one Coulomb is equal to 5.98 x 10^{-19} protons if we multiply this with the mass of proton 1.67 x 10^{-27} or 5/3 x 10^{-27}, we get 10^8, which means on Coulomb is equal to energy of 10^8 protons which is equal to one Photlon so if we slide down the Reality window by one step Photolon becomes Photon and Gravitlon becomes Magneton. But now different set of rules apply as the sub atomic particles behave as masses not energies as they do in physical world so we need to think in terms of multiples of 10^3 which is the unit mass. So Lepton is 1 x10^9, Boson is 2 x 10^18 and proton is 3 x10^27, so the mass of proton which is the side of 3 in Pythagorean triple of 3, 4 ,5 is (5/3 x 10^27), similarly we can calculate the masses of various fundamental particles.

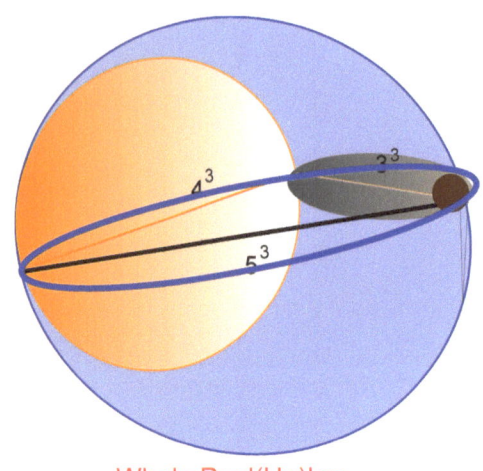

Whole Real(Un)Ion

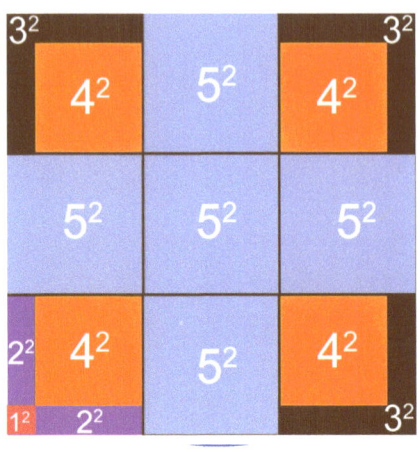

$15^2 = 9 + 6^3 = 1 \times 1^2 + 2 \times 2^2 + 3 \times 3^2 + 4 \times 4^2 + 5 \times 5^2$

Our Solar system can be visualized as the vortex formed by following the triplet sets of pythogorean cubes 3^3 + 4^3 +5^3 = 6^3 where 3^3 is the element 120 or Neutron, 4^3 is the element 121 or super proton and 5^3 is the element 122 or element Helium so should be the three dimensional periodic table.

Heim mastered the atomic world given by relation 1^3+ 2^3 + 3^3 = 6^2, so he could exactly predict the rest masses of fundamental particles as they constitute the perfect mass, Einstein mastered the fourth dimension so he mastered the gravity which is 4^3 so used 4^2 number of equations which described only the fourth dimension gravity but in fact to know the whole reality we need 5^3 or 125 equations.

Since sixth dimension is the holistic representation of all five dimensions there are five fundamental forces, not just the four known forces namely, strong, electromagnetic, weak, gravitational forces, there is one more force called Gravitophotonic force whose magnitude is given by cube root of electromagnetic force, gravitational force and the gravitoelectromagnetic force

So the matrix of fundamental particles and force carriers is of order five and not four as per the standard model. The force carrier for the fifth force is prasadon and there are eight more fundamental particles (I am no expert in QM so some of these might have been already discovered if not we have to)

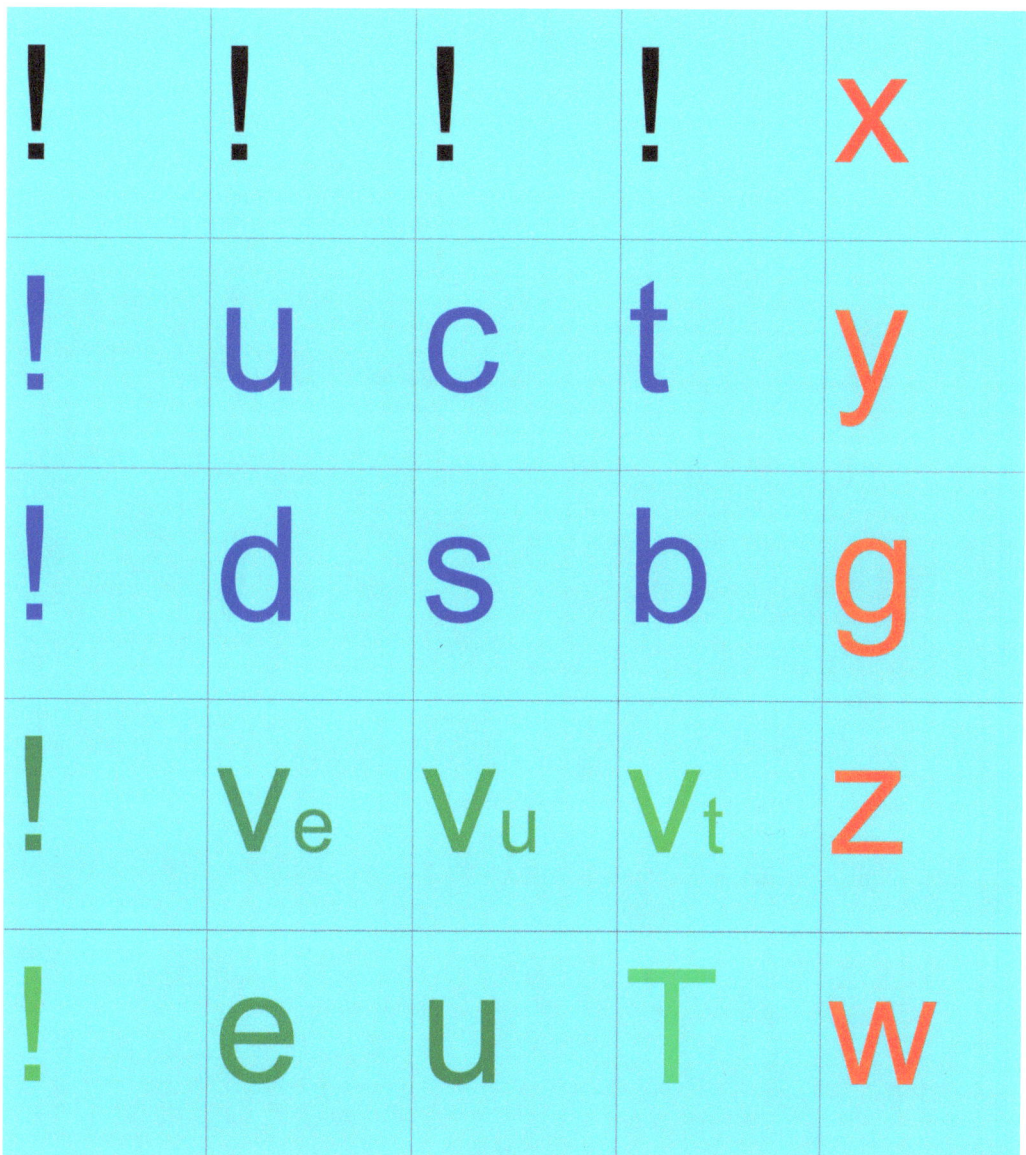

In my opinion there are as many particles in the universe as there are numbers(real including positive integers, negative integers and imaginary), of which the ones represented by prime numbers are specially stable as they cannot sub divide, so we can imagine the universe to be a sphere of numbers sprayed all over.

The quantum world involves chemistry between various particles which results in various permutations and combinations where as in the physical world we can observe finite physics

Prasad Vemulapalli

As the reality is a sliding window, we need not observe the far distant stars and galaxies to understand the universe, the miniature of the same reality is happening right under our nose in our own easily observable three worlds(atomic, physical and solar world) so we can avoid spending enormous amounts to understand the more difficult celestial phenomenon we can as well map the same to our three worlds and arrive at reasonably good results, there may be certain phenomena which cannot be mapped to our observable reference frames in which case the expenditure is justified.

Applications:

Time Dilation:

One of the most important and significant practical applications of relativity was time dilation.

Now let us see how this can be explained using theory of Reigning Element.

As I already explained the fourth energy dimension begins at C^2 and perfects at (C^2)^3.

This causes the time dilations as the velocity can vary between C^2 and C^6.

According to Lorentz Transformations,

Velocity of moving object is given by v = (1-V^2/C^2)^1/2.

If we replace V with C ,
v = (1-C^2/C^2)^1/2 = 1, which is nothing but Photlon with no mass.

So we can apply Lorentz transformations for the super world till C^3 but for super world beyond C^3, however C needs to be replaced with G thus for velocities between C^3 and G^3,

Relative Velocity of moving object is given by super v (w.r.t. R.F.3) = (1-V^2/G^2)^1/2.

Thus simply the Lorentz factor for the next R.F. w.r.t. present R.F. will be V/G.

So if we replace V with G then v = 1, which is nothing but Super Mass.

So we can retain all the physical laws and interpret things as we need by changing the R.F.s.

Conclusion: For all practical purposes space can be thought of as a sliding window comprising of one super fourth dimension, three physical dimensions and one micro dimension which can be extended in either direction (future or past) of arrow of Time., the atomic world follows closed geometry ,our world follows open geometry and the solar world (which explains the inflation theory) follows flat geometry

The universe can be thought of as a helical toroid, center of which is like a heart the artilleries of which pump out the positive energy and the veins of which pump in the dark matter which is used as fuel to produce the positive energy ,which changes polarity every few eons and flips its poles

Chapter 4
Theory of Reigning Element Vis a Vis String Theory

In first Chapter I have discussed the unification of QM and GR and now we will discuss the R.E. in relation to String Theory

Element 119 which is PhotIon has the structure of cube of dimension 0 and can be considered as point, Element 120 which is Neutron has the structure of a circle of dimension 0 and can be considered as **pi**

Now we will consider the structures of super Hydrogen or element 121, super Helium or element 122, super Lithium or element 123 and super Beryllium or element 124 of next R.F.

Please refer to the illustrations given below:

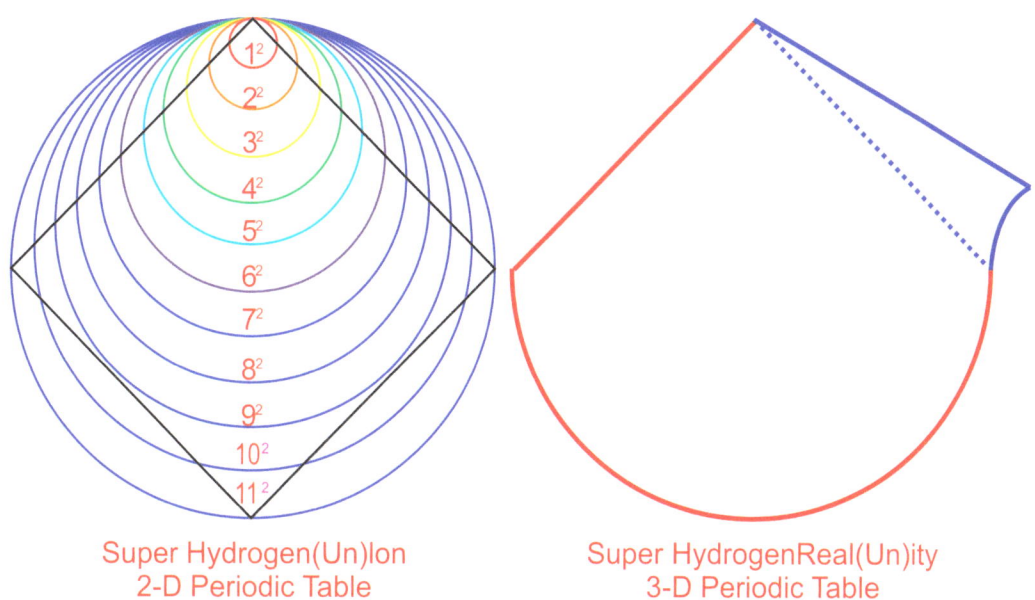

Super Hydrogen(Un)Ion
2-D Periodic Table

Super HydrogenReal(Un)ity
3-D Periodic Table

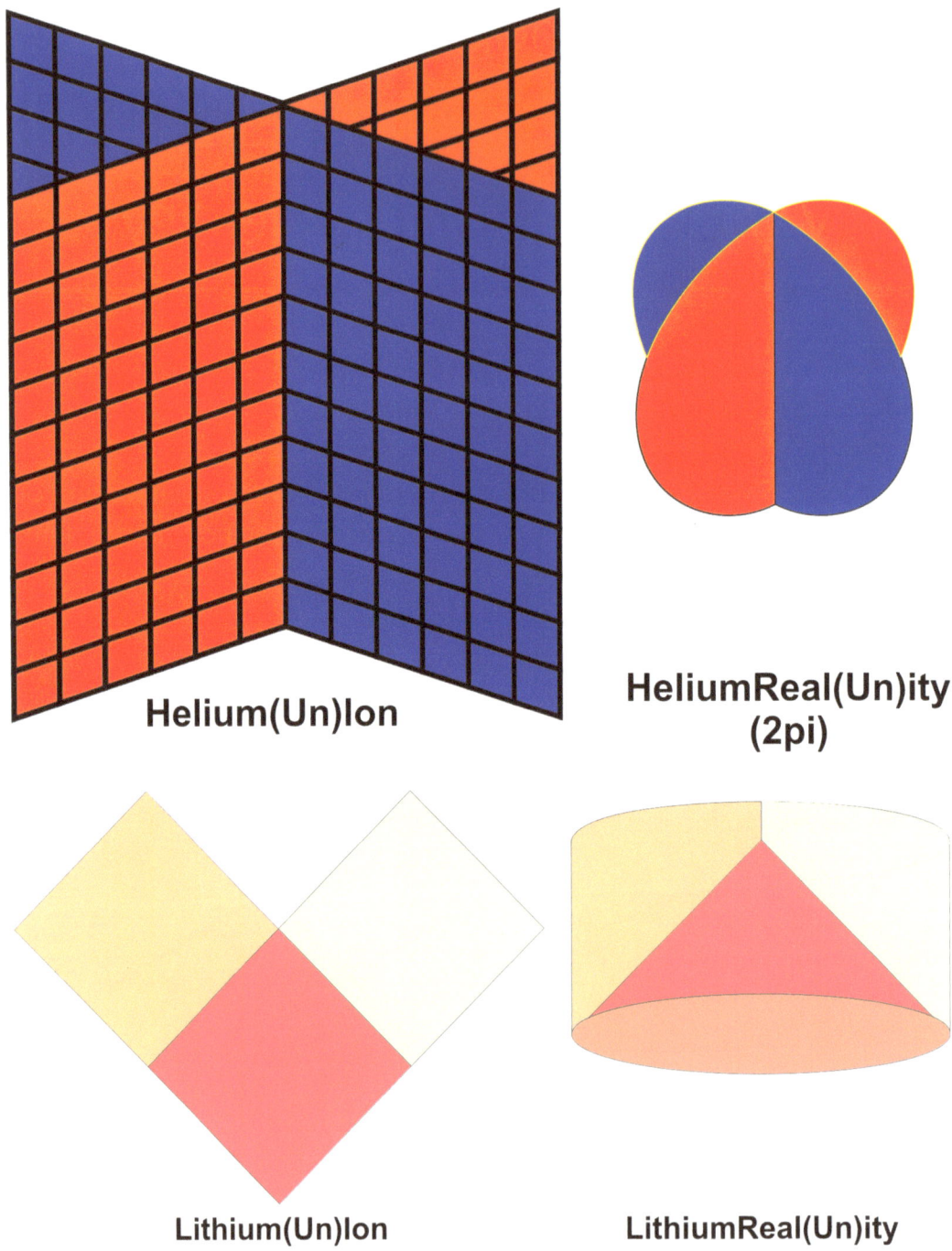

Helium(Un)Ion

HeliumReal(Un)ity (2pi)

Lithium(Un)Ion

LithiumReal(Un)ity

The hypothetical structure of element 121 is given in illustration Hydrogen(Un)Ion, that of element 122 in Helium(Un)Ion and that of element 123 in Lithium(Un)Ion

Yes, we can unify Quantum Mechanics and Relativity

The Actual structure of element 121 is given in illustration HydrogenReal(Un)ity, that of element 122 in HeliumReal(Un)ity and that of Lithium is given in LithiumReal(Un)ity

For the structure of super Beryllium which is nothing but Sun, please refer to Chapter 2

If we do the same for Helium, we get a hollow sphere, so the Helium bran is hollow sphere

If we continue with Lithium then we will have three cones (each one Hydrogen) which make one cylinder, so Lithium brane is cylindrical in shape

We know from Pythagoras theorem that $3^2 + 4^2 = 5^2$

If we take two Hydrogen(Un)Ions we get Helium(Un)Ion as shown in illustration HeliumReal(Un)ity where the vector addition of squares of 3 and 4 result in square of five and that of six and eight results in square of ten, so only seven squares namely those of one, two, five, seven, nine, ten and eleven remain, so Helium atom can be considered as **2pi**

So when we refer to a perfect circular we are referring to the atom of Helium which has 3 + 4 = 7 dimensions

We can extend this theorem to cubes as $3^3 + 4^3 + 5^3 = 6^3$

So when we refer to a perfect hollow spherical shape, we are referring to Beryllium which is perfect sphere which has 3 + 4 + 5 = 12 dimensions and can be considered as **4pi**

Element 126 is the super Carbon which is Life and is perfect full sphere and can be considered as **6pi**

We already know the shape of our solar system which is like snail shell, so the branes can be of different shapes ranging from cone as in case of Hydrogen brane to hollow spherical as for Helium brane to cylinder as for Lithium brane to spherical nail shell as for Beryllium brane so there can be 120C1 number of different branes in the universe

Periodic Table

The energy densities for the 121 elements in three dimensional periodic table are given below:

Energy	At.Wt.	Energy	At.Wt.	Energy	At.Wt.	Energy	At.Wt.	Energy
1x10	21	6x10^4	41	6x10^7	61	13x10^9	81	1x10^12
2 x10	22	7x10^4	42	7x10^7	62	14x10^9	82	2x10^12
3x10	23	8x10^4	43	8x10^7	63	15x10^9	83	3x10^12
1x10^2	24	9x10^4	44	9x10^7	64	1x10^10	84	4x10^12
2x10^2	25	1x10^5	45	10x10^7	65	2x10^10	85	5x10^12
3x10^2	26	2x10^5	46	11x10^7	66	3x10^10	86	6x10^12
4x10^2	27	3x10^5	47	12x10^7	67	4x10^10	87	7x10^12
5x10^2	28	4x10^5	48	13x10^7	68	5x10^10	88	8x10^12
1x10^3	29	5x10^5	49	1x 10^9	69	6x10^10	89	9x10^12
2x10^3	30	6x10^5	50	2x 10^9	70	7x10^10	90	10x10^12
3x10^3	31	7x10^5	51	3x 10^9	71	8x10^10	91	11x10^12
4x10^3	32	8x10^5	52	4x 10^9	72	9x10^10	92	12x10^12
5x10^3	33	9x10^5	53	5x 10^9	73	10x10^10	93	13x10^12
6x10^3	34	10x10^5	54	6x 10^9	74	11x10^10	94	14x10^12
7x10^3	35	11x10^5	55	7x 10^9	75	12x10^10	95	15x10^12
1x10^3	36	1x10^7	56	8x 10^9	76	13x10^10	96	16x10^12
2x10^4	37	2x10^7	57	9x 10^9	77	14x10^10	97	17x10^12
3x10^4	38	3x10^7	58	10x10^9	78	15x10^10	98	18x10^12
4x10^4	39	4x10^7	59	11x10^9	79	16x10^10	99	19x10^12
5x10^4	40	5x10^7	60	12x10^9	80	17x10^10	100	1x10^14
2x10^14	102	3x10^14	103	4x10^14	104	5x10^14	105	6x10^14
7x10^14	107	8x10^14	108	9x10^14	110	10x10^14	111	11x10^14
12x10^14	113	13x10^14	114	14x10^14	115	15x10^14	116	16x10^14
17x10^14	118	18x10^14	119	19x10^14	120	20x10^14	121	1x10^16

Chapter 5
Theory of Life

Theory of Life
Or
Principia Vita

Let me clarify in the very outset that the opinions expressed here under are purely intuitive and only upon finding corroborating evidence the verity can be established, so please bear with me, so far we have been trying to define life in biological and chemical terms, let us try defining life in a new perspective using physics, our cognizant predecessors gave us the following relations.

Newton proposed that Force is proportional to acceleration (dimensions MLT^{-2}, linear acceleration).

Einstein proposed that Energy is proportional to square of velocity of light(dimensions ML^2T^{-2}, areal acceleration).

Kepler proposed that Cube of the semi major axis of its orbit is proportional to the square of the period of any planet (dimensions ML^3T-2, volumetric acceleration).

Now I propose that Life L is proportional to the square of the energy(dimensions ML^4T^{-2}, volulinear acceleration). and Life as we know on earth is emitted or received in discrete packets called Vitons is given by the relation I = PV, where P is Prasad's constant and V is the vital frequency under C^2, the value of P is 11^4 x 5/3 x 10^{-27}/(3 x 10^8)^4 kg-sec^2/meter^2.

Life is the present part of Time which is the intersection of Future and Past, All life has one leg in the Future and one leg in the Past and the realities discussed in R.E.so far are true for anyone in the same R.F., so we have our perception which agrees with others too(Special Relativity physical laws are same for the same R.F., Relativity of Simultaneity), the proof for this lies in the normal temperature of Human body which is 37 Degrees Celsius, which is nothing but the heat required to for the stable element of our R.F. 3, lead 82 of which our body is made to maintain the fourth dimension i.e. 82+37 = 119 , the beginning of forth dimension. The Man as a whole is a Gravitlon, eyes are Photlons, skin is Thermlon, ears are Sonions, Nose and Mouth are Photlons thus Life, Sight, Heat, Sound and Taste are five fundamental senses of Human Body. Extra Solar Life 126 is as a whole the sixth sense.

As we have seen reality is always dual in nature, takes form of particle(bound form) as well as wave(free form), same thing happens with the Energy of fourth dimension , it takes a surprising deviation at fourth dimension, while it continues to form higher three physical dimensional objects it also tries the more intriguing route of Life, which is also four dimensional, here we need to observe one more important phenomenon, there is a continuous overlap of the energy dimensions, so we can say R.F.3 is fourth dimension of R.F.2, so Animal Life (that we can observe) starts at the beginning of fourth dimension (Protozoan) at element 82 and perfects at beginning of element 119(being the perfect fourth dimension w.r.t. R.F.3), So though there had been animal life before 119, it was purely three dimensional Life , the four dimensional Life(w.r.t. R.F.3) begins at 119, the super Ape from whom we all descended, The three dimensional Life cannot see colors and color vision starts from Apes i. e 119 ,similarly we can say the R.F.2 is the fourth dimension of the R.F.1, so Plant Life(that we can observe) starts at 66 and extends till 81, these are purely two dimensional forms of life, so this is the exact reason why plants do not have three dimensional independence, they have roots in some medium like earth, the most evolved of these plants like algae, lotus weed, water hyacinth have limited freedom of movement which enables them not to permanently attach their roots but float in water ,same reasoning can be extended to the Bacteria of R.F.0, which are purely one dimensional Life forms, which need a media to live entirely in, present form of Animal Life starts from Reigning Element 82 and extends till 126(strictly speaking from 82 to 118 it is three dimensional Life and from 119 to 126 it is four dimensional Life), by increase of each Viton a new life form takes shape, thus there are 45 main species of Life which divide into as many forms of life as they are possible combinations in 119C1, apart from 6C1 combinations of Four dimensional life(leaving our grandest father Ape).

The table of main Phyla of Life is given below.

Reigning Element	Phylum	Species
	Bacteria	
50-64	Micro Bacteria	All Bacteria
	Plants	
65-81	Plants	All Plants
	Animals	
	Invertebrates	
82-85	Protozoa To Echynodermata	Uni cell organisms like Amoeba to Star Fish
	Vertebrates	
86-91	Fish	Fishes of all varieties
92-99	Reptiles	Starting with Amphibian Frogs to Dinosaurs
100-107	Eves	Birds of all hues
108-118	Mammals	Starting with Platypus to Rodents like Mouse
	Homo Sapiens	
119	Four Dimensional Life	Ape
120	Primates	Chimpanzee, Gorilla and Orangutan
121	Caucasoid	Whites of North Europe
122	Caucasoid Aryan	Dark Whites of Southern Europe, Mediterranean and Indo Aryans
123	Semitic	Yellowish Whites or Semitics including Jews and Arabs
124	Mongoloid	Yellows of East Asia including Chinese, Koreans, Japanese
125	Australoid	Blackish Yellows or Browns including Red Indians, Dravidians of South India, Polynesians including aborigines of Australia
126	Negroid	Blacks of Africa

Therefore we can conclude that we had at least six different primates which on mutation became six of the main Human Species.

Epilogue

Incidentally according to Hindu mythology the Trinity of gods, Brahma the creator, Vishnu the Executor and Shiva the terminator rule the universe and also Adi Shiva or Parama Shiva or Parameshwara is the god of gods, we can compare Lord Brahma to Magnetion, the mass which creates all matter, Lord Vishnu to Photion which balances the needs of Magnetion and the pure energy of Sun and Lord Shiva to Gravition which undergoes fission thus terminates energy. The particle of sun which is the combination of the positive Quarks and Lepton can be thought of as the combination of Lord Shiva and Goddess Shakthi forming Adi Shiva.

This indicates that people worshipped the highest form of energy known till their time as god.

God if there is one, would be like highest dimension of energy from whom the whole world is created and is omnipresent.

Even Einstein said the same thing in other words that the past, present and the future are stubborn and persistent illusions so are the gods.

I would figure him as the guy who cranks up the great wheel of energy.

As the Unity of gods is established scientifically too, my only wish is that we the mankind should forget the differences of race, region, religion and unite as one S Union which comprises of al energy under the Sun which takes different forms as it changes into Prot(oi)on.

So I sincerely request you all not to tear him apart as he would not be pleased if you divide him to serve your monitory gains, Mercy is his currency not Money.

To conclude we are all born equal and neither are some more equal nor others any less equal

God does not gamble

References

The following concepts have been used in writing this book

I)   Special Relativity by Dr. Albert Einstein

II)   General Relativity by Dr. Albert Einstein

III)   Principia Naturalis by Sir Isaac Newton

IV)   Principia Mathematica by Sir Isaac Newton

V)   Dr. Max Planck's Quantum Mechanics

VI)   Maxwell's Electro Magnetic equations

The following books have been referred in writing this book

I)   A brief history of time by Dr. Stephen Hawking

II)   A visit to Transurania by Mir publishers

www.ingramcontent.com/pod-product-compliance
Lightning Source LLC
Chambersburg PA
CBHW050807180526
45159CB00004B/1586